# The Truth about Stereo and Video Specifications

## How to Save a Fortune When Buying Audio and Video Equipment

### Dennis Ciapura

Copyright © 2017 Dennis Ciapura

All rights reserved.

ISBN-10: 1548101443

ISBN-13: 978-1548101442

## DEDICATION

To my wife and faithful editor Diane, who has read and reread my books and articles for more years than either of us would like to admit. The adventure continues!

## ACKNOWLEDGEMENTS

The author would like to acknowledge the cooperation of the following companies in supporting his research into audio and video science over the years:

Key Broadcasting
Sudbrink Broadcasting
Greater Media, Inc.
Ford Aerospace
Triathlon Broadcasting
SFX Broadcasting
Catholic Radio Network
Broadcast Company of the Americas
Salem Media Group

## CONTENTS

|   | Acknowledgements | v |
|---|---|---|
| 1 | What This Book is About | 3 |
| 2 | How We Hear | 7 |
| 3 | The Path to High Fidelity | 15 |
| 4 | Do Tubes Really Sound Warmer? | 22 |
| 5 | Getting the Most Out of Stereo | 26 |
| 6 | Getting the Most Out of TV | 41 |
| 7 | Putting it All Together | 59 |
| 8 | Q & A | 69 |
|   | Glossary | 77 |
|   | Bibliography | 89 |
|   | Index | 92 |
|   | About the Author | 96 |

# 1 WHAT THIS BOOK IS ABOUT

If your key objective in selecting the components for your audio system, TV or entertainment center is to assemble an impressive array of premium products, then this book is not for you. If you feel that a system that sounds and looks good to you in the store will make you happy, this book is probably not for you. If, on the other hand, you are a discriminating audiophile or videophile who wants to enjoy a system that produces excellent performance without paying for technical overkill, you've come to the right place. If you are a captive of the pseudo-science, specmanship and marketing ploys you see in the magazines and feel that you're always a step behind, this is definitely a book for you. It may even change your life. This is must read book for you if you're just interested in learning more about audio science.

I've loved high performance audio and TV since I was a child in the age of vacuum tubes, and for the most part, built my own gear, sometimes from kits, but mostly from scratch. When I was old enough to grasp the intricacies of electronics, I designed my own equipment. Don't worry – this is not a book about building your own stuff!

Not surprisingly, my career path took me to radio broadcasting, which afforded me the opportunity to pursue high performance audio, but always with an eye to cost efficiency. As things developed, I was in positions to acquire or recommend equipment purchases for hundreds of radio stations and budgetary limitations were always in play. This meant it was critical to know what specifications were important in terms of what the audience would hear. I soon found that there was a definite point of diminishing returns. Every performance characteristic did not have to be perfect and to strive for perfection was wasteful and expensive.

This, in turn, led me conduct double blind listening tests, and over the years I did hundreds of them. A double blind test is when two audio or video sources are compared without the subject or the experimenter knowing which is which. Many different samples of program material are used. In many ways this is where the rubber meets the road. After all, If you can't tell the difference between the sound of two amplifiers in an instantaneous A/B test, for your hearing ability, there is no difference. This is the same kind of testing used to determine the efficacy of drugs where the drug evaluated is compared to a placebo.

I've also conducted hundreds of tests using various types of test signals to stress the systems being tested. The results of this testing sometimes suggest what to listen for in A/B testing with program sources. When audible defects are found by either test method the next task is to see what performance parameters are related to the deficiency so we know what to look for in the specifications. In all of the years of testing various equipment and systems, I've never found an audible problem or video defect that didn't show up in the performance specifications. After all this testing you begin to acquire a feel for what performance parameters are important and which are not.

There are also audible problems related to how interconnected units play with each other. A good example is speakers and amplifiers. An amplifier that sounds great with one set of speakers may easily sound stressed with another if the conversion efficiency from electrical power to acoustic power is too low for the desired listening levels and amplifier power.

As you might imagine, after years of engaging in this kind of electroacoustic sleuthing, one acquires a pretty good feeling for what it takes to present a transparent performance and what is beyond necessary. If we could wave a magic wand and obtain a degree of audio and video perfection, why not go for the ultimate everything? Unfortunately, in the real world, that's an expensive proposition.

The advertiser-supported consumer audio and video magazines will lead you to believe that there is a constant stream of performance improvements available to those who will only buy the promoted products. Unlike kitchen appliances, vacuum cleaners, etc., many folks don't replace their audio and video equipment because it has stopped working and isn't reasonably repairable. They often buy new gear for alleged improved performance. For hobbyists who have the economic means to easily do so, why not? Most of us have our little extravagancies. However, I've know people who have become so hooked on the A/V pseudo-science that they spend money that their family budget can ill afford in quest of performance levels well beyond what they can hear. In many cases, placebo effects often seem to justify these "investments" and over time a sad spiral of wasted resources can evolve.

This book will cut through the pseudo-science and present a reasonable way to approach A/V planning and purchasing. We'll start by taking a look at how we hear and what we can hear and what we cannot. Then we'll trace the history of real audio science in parallel with professional and commercial developments. For many years, designing audio equipment that could really reproduce everything humans could hear was a real adventure. Later, television went through a similar decades-long development period.

After we've explored how we got to where we are, we'll take a look at where we actually are in terms of A/V reproduction accuracy. Then we'll determine what level of performance most listeners and viewers really need to experience audio or video programming the way the programming originators intended. We'll also take a look at the effects of component placement and room conditions on performance. This will help you avoid common pitfalls and get the most out of whatever system you have.

With respect to TV, we will mainly focus on surround sound, subwoofers, center channels etc. The picture component of A/V entertainment units is mostly a function of screen size and type and screen resolution, and are thus easy for the user to evaluate. You look at the screen and what you see is what you get, but this book will help you get everything out of the set that it's capable of.

Enough detail is provided in the main text to quickly give you the information you need while the Q & A in Chapter 7 provides more in-depth background and analysis for those who really dig this stuff. So, let the adventure begin…

## 2 THE HOW WE HEAR

Understanding how we hear is critically important to understanding what it takes for electronic and acoustic systems to present an accurate replica of the original performance to the listener. First of all, it's very important to understand that hearing is a function of the signals from the ears and processing in the brain. This ear/brain system is capable of incredible processing.

For example, we are all aware of our ability to hear through a noisy restaurant or party environment to focus in on a conversation that interests us. This well-known phenomena is called "Cocktail Party Effect" The brain can actually separate the background noise from desired speech. It's not perfect, but given the fact that nature's intent was not for us to harvest gossip, but to save our butt if there is a bear about to eat us, it's great.

The brain is also capable of inserting missing information to yield a more complete sonic message.[1] A well-known demonstration of this is listening to a recording of a very familiar poem with some words edited out. You will hear the missing words, even though they are not there. Your brain dips into its memory and fills in the blanks. When you are shown the text of what's actually in the recording, you hear the skips where the missing words should have been. Another of the more amazing percepts can reveal whether you are biologically left or right handed by listening to a series of tones in headphones.[2] There are many other astonishing examples of audio percepts that are well known in the fields of psychoacoustics and professional audio.

How about those microphones that we have planted on the sides of our heads? How do our ears function? We've all seen the diagrams of the little bones leading to eardrums, so I won't bore you with another ear chart, but let's talk about how well the things work. If you're an audiophile you're probably accustomed to the notion that human hearing extends from 20 to 20,000 Hz. This might lead you to believe that the ear's sensitivity is about equal over that range, but nothing could be farther from the truth. The chart below tells the real story.[3]

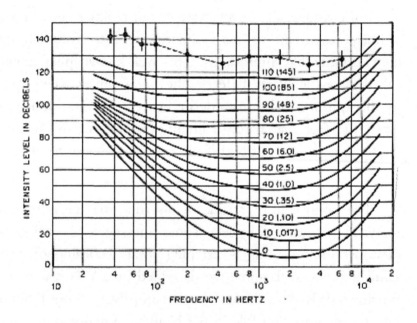

Our Auditory Response is Very Uneven

This equal loudness chart shows the frequency response of average human hearing at different loudness levels in terms of how much power is required for equal loudness. One hundred twenty decibels is near the threshold of pain and zero decibels is the threshold of hearing. Think of this as a depiction of acoustic power needed to produce low and high frequencies as loudly as the midrange output. Note the extreme difference between the high and low frequency sensitivity at low levels compared to high levels. This shows the great degree of bass and treble boost needed to produce reasonably equal perceived loudness at all frequencies. The loudness controls found on many amplifiers provide some measure of bass and treble boost as the volume is turned down. The actual response for individuals varies with sex and age. Females generally have greater high frequency sensitivity and high frequency sensitivity falls off rapidly with age for both sexes. The top line by itself shows the threshold of feeling, which is well above 100 decibels, but easily attainable by rock bands at close proximity to their speakers.

By the way, the decibel scale is not a linear measurement and 40 decibels is a loudness ratio of 100 to 1 and 60 decibels is a ratio of 1000 to 1. So, the response variations versus level, and even the at any given level, are very large indeed. Therefore, a 1 decibel response variation in an amplifier, which is about 10%, is hardly a problem. In fact, 1 decibel is generally accepted as the smallest level variation that can be perceived.

As you can see, the frequency response is depicted as 30 Hz to 15,000 Hz., which is the real limit of human hearing, not 20 Hz to 20,000 Hz. Sound below 30 Hz is subsonic rumble and few speakers other than large theater systems or excellent active subwoofers can even reproduce 30 Hz cleanly. With typical systems, what ultra-low bass aficionados are really hearing is second harmonic distortion as the speaker cones flop around struggling to couple those low frequencies to the air. Ironically, very clean low frequency speakers usually sound a bit thin in the bass because they're not generating their own false bass notes in the form of second harmonic distortion called "doubling." There are no natural musical frequencies below 30 Hz, but movie sound effects often do go that low for explosions, etc. As for the 15,000 Hz to 20,000 Hz range, few people can even hear 15,000 Hz unless it's radiated at a very high level.

Another rather shocking audio demonstration is comparing the sound of a 7,500 Hz square wave to a 7,500 Hz sinewave. This is what the two waveforms look like on an oscilloscope.

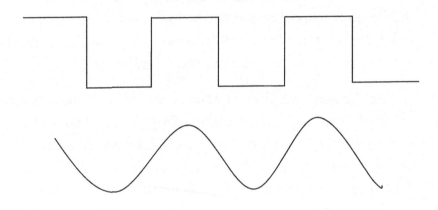

Believe it or not, these waveforms will sound exactly the same. What makes the square wave square is a stream of harmonics beyond the fundamental frequency and human ears can't hear those components above 15,000 Hz. If an electronic filter with a cutoff frequency of 15,000 Hz is connected to the square wave output of a function generator, a pure sinewave is the result. This is called "integration". The limited frequency response of our ears functions identically to an electrical low pass filter! So, what's the point of having audio system high frequency response extended to 20,000 Hz? It generally doesn't hurt anything, and if it's easy to do, as with an amplifier, it's harmless overkill and not too expensive.

## Effects of Distortion

The other key audio performance parameter is distortion, and there are two basic types of distortion: harmonic and intermodulation. Harmonic distortion is generated when the waveform to be reproduced is misshapen by harmonics of the fundamental frequency. These can be either odd or even harmonics. Even-order harmonic distortion is not as irritating to humans as odd-order because even harmonics occur naturally. Harmonics are what gives musical instruments their characteristic sounds. A guitar C sounds different than a piano C because the harmonic structure is different.

Odd-order harmonics are more irritating to listen to. Harmonic distortion is usually expressed as total harmonic distortion, THD, and 1% is generally considered to be the acceptable limit for reasonably clean audio. To stay well under that limit, 0.1% THD is a common amplifier specification and is relatively easy to achieve, even with inexpensive gear. High-end amplifiers frequently go down to .02% or less, which can be expensive overkill.

Because speakers must effect a conversion from electrical to mechanical and then acoustic output, achieving low distortion is more difficult, especially at low frequencies. If fact, distortion specifications are almost never given for speakers. In the course of my research I've measured some speakers and 1% THD in the midrange is about as good as they get. At low frequencies, 3 to 5% down to the system resonant frequency is common. At resonance and below figures of 10% or greater are not unusual.

Intermodulation distortion (IMD) is generally the most hideous form of distortion because it's not directly related to anything remotely musical. It happens when two or more frequencies actually modulate each other because of system nonlinearities. It also becomes audible at about 1%, but is really grungy. A particularly insidious variety is transient intermodulation distortion TIMD, which was a characteristic of some early integrated circuit amplifiers. In these designs high negative feedback was employed in an effort to obtain exceedingly low distortion levels and this resulted in instability. Fortunately, these designs are no longer employed.

In summary, unless something bizarre is happening in a system, if all forms of distortion are a fraction of 1% you won't hear them. What causes them and what they sound like is almost irrelevant. The point of exploring them a bit here is to illustrate that having exceedingly low distortion specs is not going to make an audible difference.

## What About Digital Compression Schemes?

Another interesting and somewhat mysterious aspect of how we hear is how digital compression schemes can be used to take advantage of what we can't hear. When digital transmission and storage were being introduced, it soon became obvious that enormous bandwidth and capacity were required to handle full fidelity audio and full resolution video. This gave rise to intense concentration on developing digital compression techniques.

There are two basic kinds of compression: lossy and lossless. Lossless compression exploits statistical redundancy analysis to eliminate duplicate data. Lossless schemes do not noticeably alter the original audio or video. Lossy compression for audio takes off the gloves and actually modifies the audio content to maximize the degree of compression. For example, a 640 MB CD can contain about 1 hour of music with no compression or 2 hours with lossless compression. However, MP3 lossy compression can record 7 hours of compressed music! This remarkable data compression is achieved using psychoacoustic manipulation techniques to alter the audio in the least noticeable ways. The algorithms employed anticipate what alterations we either can't hear or can't hear very well and eliminate data that would otherwise preserve these nuances.

Unfortunately, the mp3 and similar compression schemes are slightly audible. At first blush mp3 tracks sound great and there are no gross distortions or frequency response losses. However, if an mp3 track is compared to a CD track in A/B testing the differences can be quite noticeable. Dense rock or orchestral music with lots of instruments playing at the same time are most affected. An easy vocal with a gentle instrumental background may sound identical to the original.

Mp3 encoding is used in the majority of digital downloads and portable audio recorder/players and millions upon millions of users think they sound great. DVD movie audio is also mp3. With what we've learned about how we hear I guess it should not be surprising that the vast majority of users of compressed audio devices aren't even aware of the compression.

Most radio stations these days play their music from digital automation systems. Years ago when digital storage was very expensive, imagine the cost to record an entire radio station format digitally! So, most radio automation systems employed audio compression to bring the cost down. Today, most radio stations record without compression, but the "HD" digital broadcasting signals themselves are compressed. FM iBiquity, the FM digital broadcasting system, allows up to three digital signals along with the station's original analog signal, all on the same channel. The iBiquity digital audio is mp3.

The analog broadcast signals, which are the majority of what you hear on the air today, are not digitally compressed, but use audio analog processing that has even more audible artifacts than digital compression. Stations employ the analog processing to increase their loudness and to intentionally modify their sound for unique dial presence. This processing induces clipping, level compression, high frequency peak modification and frequency response alterations.

This overview of how we hear and what we might listen to with our high performance home entertainment system leaves us with an interesting conundrum. Why did we need a home system with 5 to 100,000 HZ frequency response, .02% THD and 100 dB signal to noise ratio?

# 3 THE PATH TO HIGH FIDELITY

In the 1950's there was a conspicuous shift from just plain radio and phonograph sound, with achievable loudness and tone being the key figures of merit, to "High Fidelity". If you were around then you will remember folks describing their favorite radio as having "a nice tone" and occasionally showing off the "volume" it could muster.

As Hi Fi sets became popular the whole jargon changed. The single knob tone control became separate bass and treble controls and Hi Fi owners boasted about the actual watts of output power their new sets possessed. Few people knew how many watts their old radios or record players put out. They delighted in turning the bass control all the way up and taking satisfaction in the thumpy sound that most of the previous generation of radios and phonographs could not produce. But Hi Fi didn't happen all at once.

The first commercial promotion of high fidelity was in the mid 1930's when premium radio manufacturers E.H. Scott and Avery Fisher marketed AM radios with exceptional fidelity.[4] Although AM broadcast frequency response extended to 10,000 Hz, the range that AM receivers reproduced was limited mostly by selectivity, which is the ability to separate stations that were close together in frequency. This had been a primary radio design challenge since the mid 1920's. Getting sharp selectivity and broad flat bandpass within the desired channel was very difficult.

Avery Fisher simply used tuned RF receivers that inherently had very poor selectivity while Scott actually developed a method of providing variable selectivity so that maximum fidelity could be obtained for varying reception conditions. Both provided audio amplifiers with excellent fidelity for the day.

These early forays into high fidelity employed AM, or amplitude modulation receivers, the only radio transmission and reception that was available in those days. But don't take for granted that the AM fidelity was as inferior as what you typically experience today. AM broadcast transmitters, even then, were capable of 1% or less THD and frequency response flat to 10,000 Hz. The difference between 10,000 Hz and 15,000 Hz response would be very difficult to perceive in double-blind A/B comparisons. Also, there were a lot more live performances than there are today, so you weren't listening to broadcasts of 78 rpm recordings all the time.

Probably the ultimate high fidelity instrument of the day was the E.H. Scott Philharmonic introduced in 1937.[5] It had the variable bandwidth tuner, an RF sensitivity control, a loudness compensated volume control, bass and treble controls, an adjustable scratch and noise filter, a 12" or 15" woofer, two 5" tweeters and 40 watts of audio with less than 1% distortion. The chassis and all major components were chrome plated! All this chrome and technology did not come cheaply. In 1937 it sold for $750, about the same as a new Ford! I am fortunate enough to have a pristine Scott Philharmonic and can tell you from first-hand experience that strong local AM broadcasts sound like FM.

By 1939 the Philharmonic was also equipped with the original FM system in the 50 MHz band. This had roughly the same fidelity as today's FM, but no stereo. All FM experimentation ceased for the duration of the War, and the current FM band was introduced after the War when the original FM radio frequencies were assigned to TV.

After decades of talking about improving the AM fidelity of receivers, which were typically flat to only 2,500 Hz with no response above 5,000 Hz, to give AM broadcasters a better opportunity to compete with FM, the Electronics Industries Association and the National Association of Broadcasters introduced an AM quality certification program called AMAX. AMAX required at least 7,500 Hz frequency response, noise blanking and other features. I had one of these radios in a 1995 Buick Riviera and it did sound like an FM receiver on strong local stations and had AM stereo as well! It also had a bandwidth switch to change to narrowband to listen to weak stations when high fidelity reception wasn't practical.

Over the years all attempts at getting the radio makers to build good fidelity AM sections in their radios failed, including AM Stereo and AMAX, as broadcasters decided to wait for digital, which finally happened in 1993 with the introduction of iBiquity, which has been slow to take off with only 35% of new car radios equipped with iBiquity by 2014.

Although it has taken the radio industry years to catch up, the science of high fidelity was making great strides over 80 years ago! On April 27, 1933, The Bell System conducted an amazing demonstration of three-channel stereo by reproducing the sound of a live symphony orchestra playing in Philadelphia at the Academy of Music for an audience in Constitution Hall in Washington, DC.[6] Special transmission lines were required to carry the audio from Philadelphia to Washington since radio transmission with the ultra-high fidelity required was not practical at the time. The amplifiers on stage in Washington were 60 watts each feeding the three very high efficiency horn loaded two-way speakers. Seventy decibels of dynamic range, a power ratio of 10,000,000 to 1, was required to reproduce the full dynamics of the orchestra with no distortion The frequency response of the overall system was flat from 40 Hz to 15,000 Hz and professional observers in Washington were astonished to hear what appeared to be a live orchestra. The musical aspects of the project were supervised by Leopold Stokowski.

By 1933 the science of human auditory perception had advanced to the point that performance specifications could be developed for everything from microphones to speakers. It was an extremely complex and ambitious project and the report documenting it is over 300 pages long. The purpose of the Bell System Philadelphia to Washington project was to demonstrate that nearly perfect audio reproduction could be achieved at the then current state-of-the-art.

As exploration into binaural vision produced stereoscopic picture viewers, experiments revealed that binaural audio also produced more realistic sound. After all, we have two ears for some pretty good reasons. The invention of modern stereophonic reproduction is usually credited to a Brit named Alan Blumleim.[7] His earliest patent was filed in 1931 and his original interest was in film audio where he saw the need for stereo's sound localization properties.

In fact, it was movie audio that drove many developments in multichannel sound, eventually leading to Dolby noise reduction being introduced in 1975. Dolby was key to saving audio cassette recording from earlier obsolescence due to the medium's relatively high noise floor. Like digital compression to come later, the key to Dolby noise reduction was manipulation of the audio in ways that were not very noticeable. In the case of Dolby, the associated medium's levels and frequency response were dynamically modified in a way that was not terribly audible. The basic scheme was to boost the levels and high frequency response of lower level program content during recording and then cutting the high frequency response and reducing the levels in a complimentary fashion on playback. Reducing the levels and treble content on playback reduced the noise level. Did you realize that all this was going on when you recorded and played back your Dolby cassettes?

## Recorded Audio

What about recorded audio? Originally developed for film and broadcast audio use, the introduction of 33 1/3 rpm LP consumer recordings in 1948 offered more recording time and a major reduction in record surface noise. Nevertheless, the disc recording system was heir to several audio impairments. From a fidelity standpoint, the major one was high frequency distortion.

The distortion on disk playback is related to the linear speed of the groove under the stylus versus the frequency recorded. This means that since the linear speed near the end of the record is about ¼ of that at the outer grooves, the distortion is 4 times greater. It is also related to the minimum stylus diameter, but that in turn is limited by a tendency of very small styli to rattle in the groove on bass tones. A partial remedy for this was the dual radius elliptical stylus. This distortion is exacerbated by the fact that the high frequencies are boosted on recording (pre-emphasis) so they can be rolled off by a complimentary function on playback to reduce noise.

Fortunately, tracing distortion is second harmonic, which is less irritating to the ear. In fact, the effect of it is to put an edge on treble content. Without the original programming to compare to the disk, tracing distortion sounds like crisper treble response.

In fact, when music CDs were first introduced some audiophiles complained that they did not have the high frequency fidelity of LPs, when is fact what they did not have was tracing distortion.

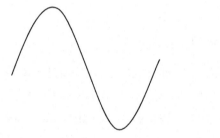 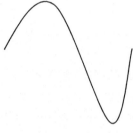

Pure tone input to cutting head        Disc playback with conical stylus

This is what high frequency tracing distortion looks like. The finite diameter of the playback stylus makes it impossible for it to accurately trace the inside surface of the very fine high frequency grooves. This is also known as pinch effect.

Eventually, recording techniques that pre-distorted the signal being recorded in anticipation of tracing distortion, thus reducing it, were developed. The RCA Dynagroove process introduced in 1963 was one such method. It worked fairly well for conventional conical playback styli, but actually produced more distortion with the newer elliptical styli, which were becoming increasingly popular. By about 1970 RCA phased out this technique.

In addition to tracing error there are distortions due to tracking error, vertical angle error, tone arm resonances, etc. so why are some people convinced that vinyl records are superior to CDs, thus driving a significant resurgence of vinyl sales? Vinyl performance is undoubtedly inferior to CDs, so it must be psychological for some folks while others just like the sound of the added distortions.

By the mid 1980's the MTS Stereo TV audio system had been introduced and this eventually led to the popularity of high performance home entertainment centers with a common stereo audio system. With TV having at last achieved something close to true stereo high fidelity audio, why not have it share the best audio system in the house?

So, after living through eight track players, cassettes, Quad and a few other "greatest" audio breakthroughs, when Sony and Phillips released the uncompressed CD format in 1982 we actually reached the peak of high fidelity for consumer use and so it remains today, although CD usage has dropped off by 50%. Mp3, AAC and other digital compression based delivery methods have taken an ever increasing share and in some ways our road to Hi Fi has taken a 180!

# 4 DO TUBES REALLY SOUND WARMER?

The dictionary tells us that an onomatopoeia is a word that looks like a sound, and "warm" tube sound is sort of like that. The hot glowing image of a vacuum tube certainly promotes the idea of warm sound. But is it really true? When solid state amplifiers first came out there was a lot of talk about "transistor sound". This arose from mainly two conditions.

First of all, the early solid state amps had limited power and it was easy to drive them into clipping. They usually stay very clean until they reach their clipping threshold with a hard clip, and rated maximum output is just below clipping. Tube amps generally have a softer approach to overload with increasing distortion until they reach overload. The maximum power output specification for tube amps is usually a little below clipping before the distortion starts to rise. Therefore, a solid state amp running at high power levels would be subject to some routine clipping while a tube amp of the same rated power would not have the same degree of clipping. Another artifact common to early solid state amps was crossover distortion.

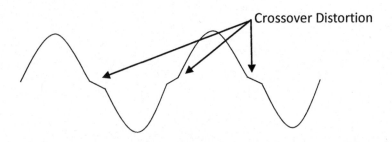

This is what crossover distortion looks like. It is the result of biasing the output stage in Class B to permit higher power output. You don't need to know what that means, just that it exists and shows up mainly at low power levels and decreases as the power is increased. Transistor power capabilities have improved so much that you would be hard pressed to find an amp today with crossover distortion.

It's interesting that clipping distortion would most likely be heard by people playing their music at very loud levels while the crossover distortion would be heard by folks operating at very low levels. It would seem that the vast majority of people listen at moderate levels most of the time where neither distortion comes into play.

It may well be that many early users of the early solid state amps were being psyched out by the magazine reviews of the time. Due to improved designs made possible by higher power transistors, neither of these distortions is a factor today. So, where is this "warmth" coming from? One way to get a warmer sound is to roll off the highs, but no amplifier of any kind has treble response that is that poor.

Transistor power amplifiers have a couple of distinct advantages over tube amps. First and most important is the elimination of the output transformers required by virtually all tube amps. Transistors have a low output impedance that allows direct connection to the speaker while tube amps require a very expensive matching transformer. In fact, the output transformers are the most expensive part of tube amps and generally the weakest fidelity component.

Transistor amplifiers with their very low output impedance

also have an advantage with respect to damping factor. Speakers typically have somewhat uneven impedance versus frequency, especially close to their resonant frequency. Impedance peaks at resonance can be several times what it is in the midrange. Varying impedance of the speaker is usually accompanied by variations in combined speaker/amplifier frequency response. So, a high damping factor tends to improve the uniformity of frequency response low frequencies. Later tube amplifier designs employing high degrees of negative feedback to reduce distortion also had improved damping factor.

## Tube Amps a Status Symbol?

With improvements in both solid state and tube amplifiers, and given the imperfect nature of human hearing, if both types of amplifier are functioning properly and operating well within their power limitations they will sound identical, and this is easily provable in double blind A/B tests. The tube power amps with their powerful looking output tubes and massive output transformers do indeed look impressive and they further a psychogenic notion of sonic superiority. And, after all, they are a serious status symbol!

There are also tube versions of preamps, control amps and even CD players, and they share some of the same halo effect. In fact, they usually manage to have their glowing tubes visible through the front panel. One of the challenges in designing low level tube amplifiers is preventing 60 Hz hum from the power supply and filament circuits from getting into the sensitive amplifier stages. Keeping the most sensitive stages as far from the power supply as possible helps, and some designs even used

DC to power the filaments, but that's expensive.

In the final analysis, There appears to be no performance differences that could account for tube amps sounding warmer. Or better or worse in any way. The bottom line is that if the user perceives any positive difference, warmth or otherwise, and is happy with the product and can afford the cost, that's all that counts. And ain't they pretty?!

# 5 GETTING THE MOST OUT OF STEREO

Fortunately, getting the most out of a stereo system does not require going out and buying the most expensive gear. Intelligently selecting each component based on its contribution to what you can actually hear and placing the speakers in the best positions available in your room can produce excellent results with a limited budget.

In terms of sound quality, the speakers make the biggest difference and selecting them should be your first step. By the way, if you doubt that the speakers make the most difference, feed a pair of good speakers with two amplifiers wired so that the speakers can be switched alternately from one to the other. The amps can be different power levels and one can be a real old one and the other the best you have.

Adjust the amplifiers for neutral tone controls, loudness compensation off and exactly equal levels. If both amplifiers are working properly and operating well within their power limits you won't be able to hear a difference when switching from one to the other. Actually, it's better that a friend do the switching so you're not biased by knowing which is the old amplifier from the garage.

The specifications you will be looking at for speaker selection are: frequency response, power handling capacity, efficiency and size. I hate to include size, but if your significant other won't let you bring your monster blasters into the house, you're dead in the water. The size is generally related to efficiency, and that will relate to the amplifier power required.

The size of the speaker enclosure is driven by what it takes to get good performance from the woofer in the system. There are dozens of speaker enclosure tuning alignments and this was a burgeoning science for years. The basic rule of thumb for years was that tuned larger boxes and horn loaded designs enable greater efficiency than small closed boxes. The gigantic Klipsch corner horn produces 105 dB of sound pressure level (spl) at 1 meter, and can produce a maximum spl of 120 dB.

On the other end of the spectrum is the venerable Acoustic Research AR-3A, which is a "bookshelf" speaker producing 85 dB spl.[8] The Klipsch corner horn produces 100 times the loudness of the AR-3A for the same amplifier power into it. One watt into the Klipsch is equal to 100 watts into the AR-3A, which is the maximum level it is rated for. The Klipsch can also take 100 watts, but will put out 100 times the acoustic power and actually reach the threshold of pain close to the speaker.

The Klipschorn is revered by audiophiles and music lovers who want to be able to reproduce the full sonic impact of a symphony orchestra in their living room. It has been in continuous production for 70 years and currently sells for about $12,000 a pair! The AR-3 was introduced in 1958, and while no longer in production, used units are highly valued. A pristine pair recently sold for $3,700! Clearly these examples of speaker technology at the extremes of design philosophy also test the extremes of affordability.

Other speaker designs employing reasonable size enclosures with efficiency about 5 dB better than the AR-3A sell for about $300-$400 per pair, and that's the range we will be working with here. In deciding your overall system budget you'll want to decide on the speakers first for a number of reasons. First of all, as stated previously, they make the biggest difference in the overall fidelity. Secondly, their efficiency will determine how much amplifier power you'll need for your room size, and lastly, you need to decide if you want to include a subwoofer, also referred to as a Low Frequency Extender, or LFE.

## The Subwoofer Cometh!

Subwoofers have become almost standard components of surround sound systems for video, but are not that popular for home stereo systems. This is a shame because inclusion of a subwoofer can enable less expensive main speakers because you won't need to purchase units that go down to the limits of low frequency audibility and they can therefore be smaller.

Consider that an excellent powered subwoofer with 100 watts and 28-200 Hz response can be purchased for as little as $100. A pair of widerange speakers that go that low would cost $900 while a pair of speakers by the same manufacturer with a low frequency cutoff around 50 Hz would cost only $300. That's a savings of $500 and a reduction in the amplifier power required!

Next is the amplifier or receiver selection. If you're planning just a stereo system now you'll want to think ahead to whether you might want to expand it to a full home theater at some point. A home theater system doesn't cost much more than a receiver without the surround sound and video accommodations, so it might be prudent to buy it now and save the expense of a receiver replacement later.

If you do go for the home theater receiver you'll find that it has a center channel output. You don't need the center channel for your stereo system unless the speakers are more than a few feet apart, like across the room, because properly phased two channel stereo creates a phantom center channel without an actual speaker. If you do use a physical center channel it should be the same model speaker as the left and right speaker systems, or at least be by the same manufacturer and have the same midrange and tweeter speakers.

When deciding on speaker placement beware of getting too much separation. Without a center channel, if the spacing is more than a few feet the sound will appear to come mostly from the speaker you are closest to, which is not stereo. A center channel speaker reduces this effect but does not completely eliminate the problem.

There is an interesting paradox with regard to corner speaker placement. It has been known since the 1930s that corner speaker placement reinforces the speaker's ability to couple to the room at low frequencies, thus significantly improving the bass response.[9] However, in a large room the corners are too far apart to support realistic stereo. Most powered subwoofers can be adjusted to produce flat response without corner reinforcement. Hence another factor in favor of including a subwoofer in your design.

After deciding whether you want a receiver or a home theater receiver, the next option to consider is the number of channels: 5.1, 6.1, 7.1, etc. At first blush, it might seem like a no brainer – just opt for the most channels in case you ever need all of them. The problem with that is you'll either be paying extra for channels you're not using or having less power per buck in your main stereo channels.

You'll also notice that receiver output power ratings are generally expressed as x watts per channel with two channels driven. This is because in modestly priced units it's not economically feasible to include a power supply large enough to power all the channels at once at full power except for the most expensive receivers. Since the vast majority of the time only the left, center and right main stereo channels will have anything playing this is a reasonable specification.

Now we get to the primary amplifier specification and that's power output. As stated previously, this is a function of speaker efficiency, room size and how loudly you like to play your program material. Most current speaker systems for the home have efficiencies between 85 and 95 dB spl, so let's use 90 dB for some approximations. For our purposes we need only consider three typical room sizes: small, medium or large

Here is the approximate amplifier power required for each of the two left and right stereo channels to produce the recommended THX reference level of 85 dB in the typical listening positions with average absorption:

| Room Size | Speaker Efficiency | Amplifier Power |
|---|---|---|
| Small 10' x 12' x 8' | 90 dB spl @ 1 m | 50 watts/ch |
| Medium 15' x 20' x 10' | 90 dB spl @ 1 m | 100 watts/ch |
| Large 20' x 30' x 10' | 90 dB spl @ 1 m | 150 watts/ch |

The above power recommendations are for music reproduction. The next chapter will deal with home theater requirements for movie audio reproduction.

With the system strategy in place and amplifier power requirements determined, it's down to the nuts and bolts features. Years ago we only had to worry about having enough analog inputs to accommodate the array of tape and disc input sources we planned to accommodate. Receivers these days have all kinds of input options, including the whole array of download and streaming services that have become so popular.

It is not unusual for a higher-end receiver or home theater system to play mp3, AAC, WMA, FLAC, ALAC, WAV, and AIFF files as well as high resolution PCM and DSD files. Many are Bluetooth equipped and some even provide Wi Fi. Some also have Sirius/XM satellite receivers. With so many options you'll need to decide which ones you need now and which you might eventually use.

If you're looking for the ultimate in audio fidelity make sure the receiver or amp you select can play high-resolution files, often referred to as HD Audio, or plan to acquire an outboard hi-res decoder. MP3 files have about 20% of the data capacity of CDs with a corresponding loss of fidelity. Hi-res capacity is three times that of CDs! There are two chicken and egg paradoxes with regard to hi-res right now.

The first is, where is the program material to come from? Simply putting CD content on hi res won't yield better fidelity than the source CD. So, true hi-res content either comes from new dubs of hi-res digital masters, such as they exist, or entirely new product created in hi-res. The second challenge is delivery. There is no hi-res streaming right now, so download is how you get the product. This means interfacing with the Internet, usually by having a computer with the downloaded files connected to the amp, receiver or outboard decoder.

## Connecting It All Together

Since we're looking at possibly connecting things, let's consider the increasingly popular Bluetooth option. Most Bluetooth encoding is lossy compression like mp3 or similar. That means less than CD fidelity and nowhere near Hi-res. In fact, since many of the music sources you will probably be using are already using digital compression, if you're a serious audiophile, you won't want to subject those to another layer of compression with Bluetooth connections to speakers, etc. The exception would be surround speakers for home theaters. There's not much playing in those channels most of the time and avoiding long runs of hard wiring to difficult-to-reach locations makes Bluetooth very attractive in some cases.

As for wiring, don't waste your money on so called "premium cables". For speaker connections, I assure you there is no difference in the "sound" of standard speaker cable with the white phasing stripes and the alleged high performance cables. Just make sure you use the correct gauge for the length of the cable runs. If in doubt, go to the next size up and you'll still be at a fraction of the premium cable cost.

The same philosophy is true for low level cables like those connecting signal sources to your amp or receiver. Good quality standard cables are all you need. If you could see the miles of cable that signals go through in recording studios and radio and television stations you would immediately question how much difference your little cable could possibly make!

## Equal Loudness Control

A front panel control that some receivers provide is an old friend from the past that has almost disappeared – a loudness control. As you will recall from the equal loudness curves shown in the last chapter, our sensitivity to low and high frequencies falls off as the loudness level decreases. These Fletcher-Munson curves, named after the researchers who introduced them in 1933, showed that bass and treble boost should be applied in increasing measure as the volume level is reduced. **ISO226 in 2003 showed general conformity with the Fletcher-Munson curves from 1933.**

There are two methods of achieving this. The simplest, and least accurate, is to provide a fixed amount of boost with a switch. The most elegant approach is to provide a volume control and a separate loudness compensation control. When playing programming at high levels the loudness control is left at its maximum position and is effectively out of the circuit. The loudness control is then used to turn the loudness down to the desired level. It actually reduces the midrange levels, which is equivalent to boosting the highs and lows in accordance with the equal loudness curves. If properly employed, the tonal balance should remain unchanged as the loudness is reduced rather than sounding thinner and duller.

At the peak of the Hi Fi days, before the stereo boom, loudness controls were quite common. With stereo amplifiers and receivers having double the complexity, the loudness controls began disappearing. There was also a purist rationale for eliminating them because the resulting response is necessarily an approximation and encouraged excessive bass and treble boost as users boosted with the tone controls in addition to using the loudness compensation. And, after all, the tone controls are available to those who really want to roll their own loudness compensation. So, it's for you to decide. If you like the convenience of having automatic loudness compensation, go for it, but you won't have many receivers to choose from. On the other hand, if you play your music loudly most of the time you can always just use the tone controls on the odd occasions when your setting up for background music.

To summarize the equipment selection process, here's a list of things to bear in mind:

Speakers -Placement and efficiency

Amplifier Power – Adequate for speakers and room

Subwoofer – Does receiver have subwoofer output?

Center Channel – Does receiver have center output?

Digital – Can receiver decode the types of files you want?

Loudness Control – Do you want one, what type?

By the way, if you still play vinyl be sure the receiver you're interested in has phono inputs with a preamp behind them, unless you have a great external preamp you want to use.

For years the scenario for buying a stereo system went something like this. The buyer would listen to a selection of amps or receivers in his or her price range and select a pair, probably the most expensive the budget would allow. The speakers were selected in much the same way. Any listening tests were done in the store, which tells little about how the system would sound in your listening room. Next step, lug the boxes home.

Once at home the receiver would probably go in the same place as the old one, if there was one, as would the speakers, unless there was a place where the speakers would look better. Sound familiar? This venerable approach produces stereo by luck. A better method would be to analyze the listening room first because this will determine the amplifier power required, the speaker efficiency and speaker placement options.

The first room consideration is size because this relates to amplifier power and speaker placement. A small size room should allow corner placement of the speakers without losing stereo imaging. A medium size room would take twice the power and a center channel if the speaker separation is more than the distance from the center of the listening position to either speaker. This would mean selecting a home theater receiver to get the center channel output. This would also give you a preamp output to drive a powered subwoofer. We'll address speaker placement in more detail after deciding on which speakers to employ.

If you're an experienced audiophile you can probably shop the catalogs and internet to select the best receiver value, but speakers are another matter. If at all possible, at least listen to the speakers live in a showroom before purchasing them. If you're working with a really good dealer, he or she may even let you take the demo pair of the speakers you're considering home for the evening to try out in your listening environment. The idea is that you would pick them up when the store closes and bring them back in the morning at opening. This could be done on either a rental or deposit basis. Another alternative is a solid 10 day money back guarantee. This is not as good as a loaner because it would be too easy just to keep the speakers.

Now that a general plan has been developed and the equipment selected, let's take a look at speaker placement. There are tons of articles on the web about speaker placement, and not all sources give the same advice. One of the popular schemes has the speakers out in the middle of the floor with the listening area a few feet behind. This may be the optimum setup for minimizing standing waves, but it assumes that you start with an empty room and speaker placement is the prime objective. This would hardly work in most real world households.

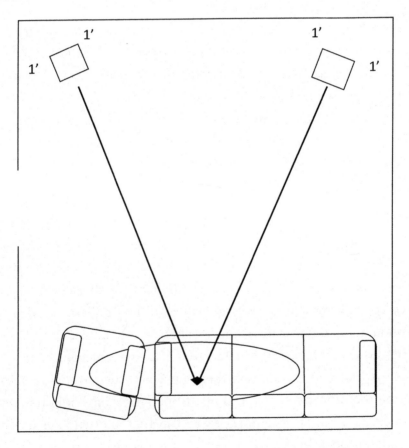

Our approach will be to recognize what's doable in most homes and suggest speaker placements that are logical and I know work from personal experience. We'll first look at a small to medium size room that is typical of most family rooms This setup is specifically for stereo applications and not intended for home theater setups, which are covered in detail in the next chapter. Above is a plan that works well for a small to medium room. The room shown is about 12' x 13'.

Speaker placement diagrams are typically shown with only a couch to designate the listening area? Who's family room has only a couch in the listening area. Even a small family room will have a chair along with the couch to provide more seating. The speakers are placed near the corners to take advantage of the low frequency reinforcement they provide, and for a music system, usually eliminates the need for a subwoofer. However, rather than being tucked tight into the corners they are pulled out about 1' from the back and side walls to reduce standing waves from reflections from the wall surfaces close to the speakers. Unless they are floor-standing units, the speakers should be on stands to bring the midrange and tweeter units to about eye level. Unless they already have their mid and high units toed in (very rare) the speakers should be toed in to focus on the prime listening seat.

If at all possible, place the speakers and listening position seating on the walls without a doorway. This preserves the wall surfaces extending from corners with the speakers so with the floor surface they can form a virtual large-mouth horn supporting the low frequencies.

The ellipse shows the approximate area where good stereo

would be produced. The prime position is the first seat on the couch (probably also the most comfortable and frequently used) but the center couch seat and chair should also receive decent stereo imaging. The right seat on the couch is reserved for the dog.

This arrangement differs a little from the classic approach of having speakers and listening position form a triangle with equal distance to the listening position and spacing between the speakers. This eliminates having furniture sitting out near the middle of the floor, which is a huge waste of space, and the stereo imaging is still excellent with no need for center channel fill. Overall, it's a nice simple arrangement for a small room accommodating a reasonably priced stereo system.

As for room treatments, small rooms usually aren't much of a problem because they tend to be more cluttered and this renders them less echoic. The increasing popularity of wood floors is a bit of a problem compared to carpeted floors, but this is easily reduced by adding an area rug in the center of the room. Hanging a few sound absorbing panels won't make a noticeable difference. If anything, small rooms tend to be a little dead rather than excessively live.

While we're talking about room characteristics, let's take a look at room equalization. In a phrase, don't attempt it. The ear-brain system can hear through the acoustics of the room and any EQ above about 200 Hz that you apply will be perceived as irregularities in the speaker response.[10] Here's how it works.

Remember the cocktail party effect we discussed where we

are able to hear a desired conversation through the noise of a busy dining room? The same ability is in play when you have difficulty making out all the words of a speaker in a large echoic space like a church or meeting hall, but after a while the intelligibility seems to improve. Your brain is learning the room! The same thing happens when you attempt to equalize a room. As your brain becomes accustomed to the "sound" of the room it perceives the response as being the sum of the speaker response and the EQ that was added.

Movie theaters are EQed for specific reasons. There is a screen loss that attenuates the highs coming from the speakers behind the screen. There is also a natural high frequency path loss from the screen to the prime viewing area part of the way back in the theater. The theater's technicians or consultants who tune the theater intend for the EQ they apply to be perceived as additive to the speaker response.

An exception would be low frequency EQ to tame room nodes that may cause unwanted low frequency peaks or dips in the listening position. If it happens that your room is causing boomy bass right in the prime listening position, equalizing this down would not affect the mids and highs where most of the program content lives.

Low frequency sound waves have such long wavelengths as to be virtually non-directional, which means that if you have a subwoofer it can be placed almost anywhere that is at least a few feet from the listening position. So, if you have boomy bass in the listening position try moving the subwoofer before resorting to EQ. If you're not using a subwoofer, try moving the left and right speakers a little farther from the room corners.

Now that you've made intelligent equipment design decisions based on your personal objectives and the listening room environment, and have placed your speakers to get the best performance they can deliver, select some of your favorite music samples and enjoy!

---

# 6 GETTING THE MOST OUT OF TV

It really wasn't that long ago that a TV was a wooden box with a long glass cathode ray tube inside flickering with a black and white simulation of a picture mired in "ghosts". We could never have imagined the gigantic high definition LED flat screens, digital TV receivers, cable and satellite integrated video storage with hundreds of hours of recording capacity and HDTV Blu-ray movie disk players that are the centerpieces of today's home entertainment centers. The good news is that there is tremendous entertainment value available to consumers. The bad news is that a lot of money can be involved and it's easy to spend too much.

The design of an entertainment center for the home usually starts with the selection of the TV, and this can go two ways. Some folks are inclined to just go out and buy the biggest screen they can afford. Others have a place in their room in mind and proceed to select a screen size appropriate for that location. Neither approach is right or wrong, it just depends upon the circumstances. If accommodating as many buddies as possible to enjoy sports broadcasts is the goal, the biggest screen approach is perfectly rationale, unless multiple screens are part of the plan. Picking a compatible size for a strategically selected location also makes a lot of sense.

As for locations, one of the most common placements is also potentially the worst: over the fireplace. It's a natural trap to fall into because the fireplace and TV naturally want to share what is usually the prime focal point of the room. The first audio option, speakers on the hearth, is better, but doesn't look very good.

A much better option from a technical perspective is to mount the flat screen on either side of the fireplace on the wall. Ideally, the seating arrangement should be shifted or aimed at the TV. The screen can be a big as required to provide an impressive presentation from the seating area and it can easily be positioned at eye level. An equipment cabinet can be located under the TV and the main speakers for a surround sound system can be located to the left and right of the equipment cabinet. Also, locating the TV off the fireplace creates a bit of asymmetry that many designers prefer. Here's an arrangement that works very well:

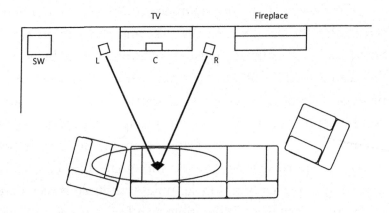

In this arrangement a complete home entertainment center can be accommodated on the wall to the left of the fireplace. The cabinet for the electronics and software is below the wall-mounted flat screen TV. The left and right speakers are floor-standing units to put the midrange and treble at about ear level and the center speaker is a bookshelf unit by the same manufacturer with matching midrange and treble speakers. The corner placement of the subwoofer improves its coupling to the room for low bass reinforcement.

The ellipse shows the area that would receive a good stereo image and reasonably on-axis TV viewing, which includes three of the five seating positions. Surround speakers are assumed to be in the ceiling. There is really very little content sent to the surround, except for a few seconds here and there, and it doesn't make sense to spend a fortune for those speakers. Take a hint from what you see in the movie theaters. Just be careful to get the hookup polarities correct so they are properly phased.

## The TV Display Options

Before we take a detailed look at setting up the audio, let's review the TV options to see where we might save some money. Today the options for home entertainment-size flat screens (45"- 65") are almost all 4K HD with HDR. The 4K part of the HD basically quadruples the screen resolution, but you need 4K signal sources to actually see the improved HD. The HDR is high dynamic range, which refers to the range between maximum and minimum screen brightness. An ideal display would go completely to black from a full brightness picture, thus making the picture a higher fidelity reproduction of the input signal.

In addition to standard 4K Ultra HD sets, we also have a new breed of OLED TVs. These sets actually allow controlling brightness at the pixel level and achieve the ideal full black. The catch is that a standard 55" 4K set is about $1,000 while an OLED set is about double that. You'll want to look at samples of the same screen size by the same manufacturer to judge for yourself whether you can really see the difference between these screen options. With today's flat screen Ultra HD TVs the picture quality of all of them is so good that it's not surprising that many folks can't see small differences that videophiles are agog about. Picture quality from the user's perspective is, after all, highly subjective, and the difference in cost between regular 4K Ultra HD and OLED could pay for your receiver and speakers!

Consumer 4K is an offshoot of 4K professional where the big advantage of the Ultra HD resolution is in the editing. For example a director may want to edit into a screen space that is a fourth of the area that was shot. With regular HD this would take that scene out of HD resolution. With 4K it would still be HD.

The newest wrinkle in displays is curved screens, which except for reducing screen refection's from side light sources is pretty much an appearance item. Surprisingly, these don't cost that much more than standard 4K Ultra HD sets.

If you're interested in watching 4K sources, the next feature you need to consider is the TV's connectivity with those sources, which at this point are streamed services like Netflix, Amazon and You Tube. You'll need at least 20 Mbps internet service to support these streamers. If you have DirecTV with the latest Genie DVR and Mini Genie converters you're all set for their 4K programming and there's at least one 4K Blu-ray player (Samsung) on the market at this writing.

What about projection TV vs. large LED flat screens? Most of the folks who have good quality projection setups love them and would not switch. Quite simply, it takes a screen several feet across to fill the viewer's field of vision at 12 feet or so. If you're the type of movie viewer who sits just a few rows back in a theater, having that full field view at home is awesome. With today's HD and Ultra HD 4K projectors the resolution of the projected image is good enough that you won't see the pixels. However, you will have to have a darkened room to keep the reflected image from getting washed out even though the projector specs for black level and contrast may be as good as a flat screen. Getting a real optically correct screen will help with this. Also, be aware there is difference between real 4K projection (very expensive) and 4K input compatible. And don't be tempted to just paint the wall white. It may look good in the abstract, but compared to a real screen it wastes image power.

The economics are another consideration. Unless you pay a premium price for an LED projector, you'll need to budget for new projector lamps every couple of years and they cost a couple of hundred dollars. That could buy a small flat screen and some folks come to resent having that operating expense. After the first lamp replacement, letting the projection TV play when no one is watching it for a while will become a thing of the past!

Well, now that you've decided on the best TV type and specs, the real fun begins. When early TVs limped along with mono sound through 5" speakers for table model sets, or perhaps a console with a 12' speaker, but with an open back cabinet and no tuning, who could have imagined today's surround systems! Audio fidelity was a radio focus and TV sound was a secondary consideration. With the frequent buzzing of the crudely designed vertical sweep circuits infiltrating the audio, the poor fidelity was probably a blessing!

Today's high performance surround sound audio systems with 100 watt powered subwoofers and 200 watt receivers often contribute more excitement to a movie presentation than the picture. The high performance multi-channel sound systems in the best movie theaters taught us what was possible and we wanted it. We might not be able to match the visual impact of a big commercial theater screen, but, properly set up, our 200-600 watt audio systems in the average family room volume can match the loudness and fidelity of the big theater systems.

With all of the talk of subwoofers and surround options, you may be surprised to learn that the most important speaker is the center channel unit. Think about it. Almost all TV dialog is mixed to center, as is a large portion of the music content. A good center channel speaker with "voicing" matched to the left and right speakers is likely to cost more than the subwoofer, but it's worth it. When a car, train, or other sound effect moves across the screen it's tonality should not change. When it does, there is an audible disconnect as the effect gets to the edge of the screen.

Once you've determined the best room setup for your listening environment (with compromises usually) it's time to play some movies and watch and listen critically. The two most common audio maladjustments are excessive bass and excessive surround. Your home entertainment center should sound realistic, and not like the bass assault in the back seat of a low rider, and the surround should not make your living room sound like an echo chamber.

Adjusting the subwoofer properly can be a bit complex. First of all, determine what kind of subwoofer output your receiver provides. Some have a fixed cutoff at about 200 Hz while others have several options going down to below 50 Hz. For most main speakers 80 Hz is a good crossover from the main speakers to the subwoofer because it's high enough to keep the low lows going to the sub, which is designed to handle those frequencies with extended response and low distortion. It's also low enough to filter out higher frequencies that might enable localization of the subwoofer.

You should never be able to actually hear the subwoofer in the listening position. If properly set up, it's low frequency contribution should be seamlessly blended with the output of the main speakers. Eighty Hz is also high enough to protect the main speakers from having to handle frequencies near their low limit. Some receivers give you cutoff frequencies based on your main speaker size - small or large for example. This can be misleading because to relatively small floor-standing speakers are almost flat down to 75 Hz, and should have a crossover in the large category. If your subwoofer has its own crossover, set it at or above the receiver's crossover setting. You don't want a hole between the main speaker and subwoofer response.

If your subwoofer has a phase switch, you want to check it next. If you've paid attention to the polarity of the wiring connecting subwoofer the $0^0$ should be correct. To be sure, play some music with regular bass content and have a helper alternately switch between $0^0$ and $180^0$ and observe which setting results in more bass — that's the correct one.

The next step is setting the subwoofer level and without some test equipment that can be challenging. Some receivers have a speaker setup program and these can range from very good to very bad. If your receiver has one, give it a try and see how it sounds in extensive listening tests. If the resulting sound is natural without excessive or boomy bass keep it.

Now the tough one. Manually adjusting the bass level without test equipment is a gray area for most users. In fact many users set the level control to mid-position and never touch it again. They accept whatever the resulting overall response is. Sometimes this works out just fine and other times it can be way off. Classical and jazz musicians who listen to a lot of live instruments can usually get a satisfactory balance by ear. Rock musicians tend to adjust for way too much bass.

If your receiver has a pink noise generator that can feed each channel individually, a good match can be achieved by adjusting the subwoofer level to be about the same loudness as the left and right channels. Follow up by playing lots of music and video programming material and fine tune the subwoofer level if necessary.

So, what does a properly balanced subwoofer sound like? Most of the time you should not hear it doing anything from the listening position, however, when the program material has honest-to-goodness real low bass it will be easily audible. A good test cut is Celine Dion's *A New Day Has Come*. The very low content starts at 2:10 min. If you don't have the CD, the mp3 version does reproduce this passage faithfully. If your subwoofer level is correct, the low bass should play about as loudly as the vocal.

When your subwoofer level is correctly set you'll notice another interesting surprise. Realizing that more and more viewers have TV systems with subwoofers, reality TV producers are increasingly using background music tracks with really low bass segments. They make a surprising appearance because you're concentrating on the dialogue and then start hearing almost subaudible sounds. For an instant you think it might be some earth quake thingy from outside until you pick up on the regular rhythm.

So, if one subwoofer is great, how about two?! You might think that the main advantage would be twice the low bass power, but it doesn't really work that way. The ear's perception of loudness is logarithmic and not linear, and adding a subwoofer would increase the acoustic power output by 3 dB and it would take a 6 dB increase in sound pressure level to double the loudness. However, there are other advantages to dual subwoofer operation.

Although you won't double the low frequency sound pressure, you will have 3 dB of additional headroom, which is protection against overload. The distortion will also be less because the speakers will be running at half the power for the same acoustic loudness. Also, the intermingling of the refection nodes resulting from having two speakers in different locations will generally result in smoother response. In other words, a dip in response of one speaker is likely to be at least partially filled in by radiation from the other.

Here's an example of a modestly priced compact home theater system that produces magnificent performance:

The TV is a 55" Samsung 6000 Series that costs about $1000. The receiver is a 10 year old 6.1 Sony with 100 watts per channel at .09% distortion. This unit was on hand and was therefore 0 cost. The left and right speakers are Boston Acoustics A25s that cost $134 each. The all-important center channel is a Boston Acoustic 225 with midrange and treble drivers matching the A25s and costs $249. The two subwoofers are Sony 100 watt SA-W250s that cost less than $100 each. The

Blu-ray DVD player is a Samsung BD-5300 costing about $100.

At first blush, this may seem like an odd assortment of equipment for a high performance system, but let's take a look at the strategy. The TV is a good solid performer and the improvement offered by a more expensive alternative would not be appreciated at the distance to the screen for which this setup was designed. Using a ten year old receiver for the audio centerpiece might also seem odd, but it provides all of the inputs and outputs needed, adjustable subwoofer crossover parameters, and has more than adequate power with negligible distortion. It sounds as good as any current high quality receiver.

It is also interesting that the center channel speaker costs as much as the left and right main speakers together and more than both subwoofers together. This reflects the reality that more of the total audio from any source is centered in the soundstage. Also, the mid and high drivers of the left, right and center should be the same and the overall speaker systems "voiced" for equal sound character.

Another reason for selecting the Boston Acoustics units is that they go down to about 50 Hz, so there is good overlap with the subwoofers crossing over at 80 Hz to ensure there is not a hole. The three surround speakers in this system are 8 inch ceiling mounted coaxial units costing about $50 each.

So, the total cost of this system is under $2,000 with the investment about equally spent on the video and audio. Despite this relatively low cost the system performance is impressive. It can achieve threshold-of-pain loudness levels anywhere in the listening room. The low frequency response is smooth and extends down to frequencies you feel rather than hear, and the purity of reproduction is amazing. By upgrading the TV and receiver and spending more for the speakers one could easily spend twice the price of this system with no noticeable improvement in audio quality. By the way, the screen is suspended on a Touchstone Whisperlift ($450, not included in A/V price) and disappears behind the cabinet.

## Creating a Super System

Now let's take a look at what our cost savings options are for a super system in a large room. The first spec we need to consider is amplifier power level. Let's say your room is 20' x 30' with a 10' ceiling and average absorption characteristics. In rooms that large most have average absorption. Unusual absorption (or reflection) would come from obvious things like sparse or dense furnishings, lots of mirrored surfaces, floor to ceiling windows along more than one wall or textured wall coverings.

## Amplifier Power

From our power vs. room size chart on page 30 we see that 150 watts per channel in a stereo system would be ideal. However, a home theater receiver with 100 watts into left, center, and right for speakers averaging 90 dB per watt sensitivity is more than adequate. In fact, receivers with 150 watts per channel are extremely hard to get and very expensive. Allow a little more if your room is unusually absorptive and you like to listen at the threshold of pain, and feel free to cut back to 100 watts if it's unusually reflective. You are very unlikely to be working with speakers varying much from 90 dB unless you choose horn-loaded systems, in which case almost any amplifier power down to 10 watts will work fine.

At this point let's review how amplifier power is being specified these days. The 100 watt class amps may be rated at 90, 80 or even 70 watts to get the THD figures below 0.1% They're all really 100 watt amplifiers, and in fact, may be the same designs. So, think terms of power classes rather than thinking there's some advantage of getting a few extra watts.

The other interesting specification is the number of channels driven at the rated power level. With as many as nine amplifiers available (7.2 surround and subwoofers), if all were driven to full power at the same time the load on the power supply would be enormous and expensive. Given the fact that the front two or three channels will be the only ones playing more than a fraction of a watt most of the time, the two channel stipulation makes a lot of sense.

So now that we know the power class to shop for, our receiver selection job becomes a matter of getting the features we need or think we might need in the future. Since we're shopping for the heart of a super system, give this list some careful thought. At minimum, you want to be sure that all inputs provide 192kB/24 bit digital to analog conversion and that Dolby and DTS surround formats are decoded. If you're planning ceiling speakers, you may also want to check out Dolby ATMOS, which enhances the sound field of ceiling speakers.

Now the first step in keeping the cost of our super system out of the super cost realm. Dealer open box or scratch & dent sales are great sources of new equipment at low prices, and as you probably know, scratch & dent merchandise is rarely scratched or dented. More often than not is was returned by a buyer who didn't know how to use the gear and it can no longer be sold as new. Sales of last year's merchandise can also be very attractive. The customer service reps at major retailers and on-line outlets tend to be very helpful. Used equipment direct from consumers, on the other hand, can be a bit risky. You don't know if the equipment has been abused, like having been operated in an unventilated cabinet at high temperatures.

## Selecting the Speakers

Now it's time to move on to the speakers, and for the super system theses have the most influence on the sound quality. The first decision is whether you want to go for powered systems. Good powered systems sell for about $3,000 and the question is what you get for the money. You probably will already have 100 watts per channel in your receiver which will become redundant if you have powered speakers. The only advantage to powered units is that they can be fed wirelessly. For a super system I'd rather see the money go into better speakers or equivalent speakers with the savings going toward other sonic upgrades, like high resolution audio.

As pointed out previously, it's important that the characteristics of the center channel speaker be identical to that of the left and right front channels. There are a few center channel speakers that are "voiced" to match the company's left and right units, but in most cases you're on your own. The safest approach is to buy three identical floor-standing speaker systems. But which ones? Since we'll plan to use a subwoofer, we don't need the extended bass response that the taller and more expensive units offer. The shorter center channel speaker also avoids having a speaker system close to the TV screen which is a distraction.

In many ways, the subwoofer selection is the most difficult. They all look about the same, they all have very similar specs and since they all have bass level controls it's very difficult to tell the difference in their sound.

For our super system, I recommend dual subwoofers for several reasons. First of all, since the room is large, dispersing the low frequency radiation will minimize the likelihood of bass hot spots near a subwoofer due to proximity. Also, two subwoofers of lesser price are likely to have more output capability than one more expensive one, even though the more expensive one may go a couple hertz lower at the -3dB point.

The surround speakers are way overdone in many home theater systems and often set to play too loudly. Take a lesson from the movie theaters. The surround speakers are very modest compared to the monsters of audio technology behind the screen. There is a reason for this.

The surround audio is normally both very occasional and very brief. The surround effects sent to the surround speakers also tend to be mostly midrange. All low frequency components are crossed over to the woofers and subwoofers. For home systems, good quality ceiling speakers do the trick, unless you have your heart set on having surround effects jump out at you in a way the producers and directors never intended.

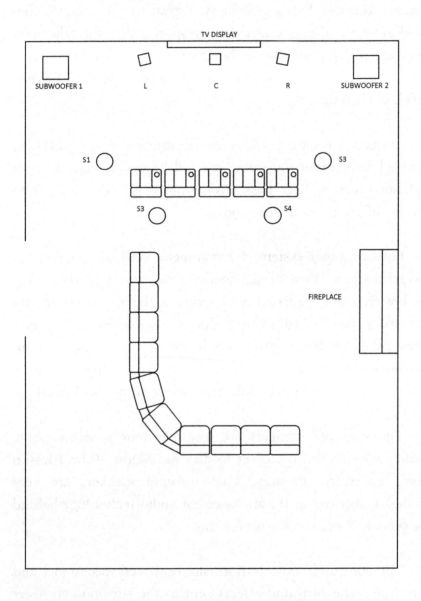

This being the case, ceiling surround speakers rather than floor-standing units will also save a lot of money, including wireless facilities to avoid some very nasty speaker cable challenges.

The plan on the previous page illustrates one possible large room layout accommodating a super system. The anticipated system consists of a wall-mounted display with a 7.2 audio. The four surround speakers are in the ceiling.

The 30' x 40' room is divided into a home theater area and a conversation area dominated by the fireplace. A flat screen display is depicted, but a large screen for an overhead projector would work just as well with the same speaker placements. The only possible difference would be to move the theater seats a few feet farther back until the screen just encompasses the field of view from the center seating position. In other words, you see the entire screen width without moving your head. If you have to move your head to take it all in you're too close. If you see the screen and the things surrounding it you're too far away. Your screen is your window on the world!

The subwoofers are shown in the preferred corner placements. If the bass is too dominant turn the speaker levels down before moving out of the corners. The corner placement gives some degree of low frequency extension in addition to overall bass gain, so you don't want to lose that benefit.

The floorstanding front speakers are shown in the proper positions for the flat screen and would be spaced toward the edges of a projection screen to match the wider field of view. The ceiling surround speakers are for a 7.2 system that includes side and rear surrounds. If you use floorstanding surrounds, place them a couple of feet farther from the listening position.

## So What Does it All Cost?

So how much would you spend on this super system? With careful shopping, an excellent receiver should cost about $300; the main speakers about $600; the subwoofers $300 and the surround speakers $200. This makes the audio total a very reasonable $1,400. The video can be almost anything you desire, starting from about $1,000 for a 65" 4K flat screen to $16,000 for a real 4K native resolution projector with long life LED bulbs and a 150" real HD screen. A more reasonable maximum would be a good quality 4K compatible HD projector and the same screen for about $1,600 total for the video and $3,000 for the system.

---

# 7 PUTTING IT ALL TOGETHER

If assembling a stereo or home entertainment system is taken as a scientific pursuit with a tinge of love and excitement, it can be a very rewarding adventure. The anticipation of turning it up and hearing your favorite pieces of music or watching your favorite movies can be thrilling indeed, especially as you hear and see things you never knew were there before. That's a true revelation. And even more rewarding is you know all of your investment has been in system performance that you can really see and hear.

I've made a real effort to save you from the expense of pseudoscience overkill while sharing my four decades of audio and broadcasting experience as it can be applied to home entertainment projects. Some may disagree with some of the advice given herein, and that's OK. The best I can do is share what I believe to be correct and true. Equipment sales folks probably don't think much of my approach to equipment purchasing and advocates of audio and video mythology definitely don't subscribe to my view of their parallel universe.

Audio and video have made such enormous strides in introducing new kinds of gear that we could never have imagined a few years ago, that it's easy to assume that every new gizmo or improved product type is really better than what you already have or can get at a terrific discount. But it pays to be a little skeptical.

The first key to productive analysis is understanding what you really can see and hear so you know what specifications you need as opposed to what the magazine review and promotional materials tell you that you must have. Hence the HOW WE HEAR chapter. If you don't completely understand what's presented here, it's worthwhile to read it again. Further research on the Web may expand your knowledge, but at the risk of picking up some really atrocious misinformation. It's like anything else on the Web. Almost nothing is peer reviewed and webquackery abounds.

## Getting to Info Quickly

You may find the Q & A in the next chapter helpful in filling in some details not covered in the other chapters. You may also find the INDEX and GLOSSARY useful if you saw something of interest and need help in getting back to it.

Now that you've put your system together, it's time for the real fun part: testing and adjustment. The first logical step is to make sure that all of the audio channel levels are the same. Most receivers and amps have a built-in pink noise generator that rotates around the channels to verify levels. In addition to listening for level variations, it's also a good idea to listen to the tonal quality of each speaker system.

## Checking Levels and Phase

The three main channel speakers should sound the same. If one of the channels lacks high end, put your ear up to the tweeter to confirm (or not) that the speaker is playing. If one of the channels sizzles but has lower volume, check the midrange speaker to see if it's playing.

The subwoofer should sound like a low rumble with no noticeable mids or highs. This is also a good time to check your subwoofer phasing. Change the phasing selector while the subwoofer is playing the test signal and see which position produces the loudest output. Leave the phasing switch in that position and reset the subwoofer levels if necessary to match the other channels.

## Adding the Subjective Element

In the very beginning, we recognized that audio reproduction is both science and art. In starting out with proper equipment selection, the ideal placement of speakers and attending to the level and LFE phase adjustments, we've covered most of the science. Now we need to see if we can improve the performance of the system with some personal preference alterations.

Given that our super system has subwoofers, the first subjective assessment should be the bass volume level. Even though you may have set the starting levels to match the other speakers with the receiver pink noise generator, after listening to some program material you may feel that the low bass is too thin or too dominant. In either case, readjusting the subwoofer levels should solve the problem. However, if the low bass is OK, but the mid bass is boomy, you may want to experiment with speaker placement. The corner placement we started with should give a bias to low bass, but if that's not what you have, then move them in or out of the corners about a foot at a time. If that doesn't work, try putting them flat against the walls and a few feet away from the corners.

The next levels you may want to review are the surround levels. This will take a lot of listening to a lot of movies with some surround. Beware of setting the surrounds too high as this may sound exciting at first, but long-term will sound unnatural and irritating. It is <u>very</u> useful to pay particular attention to surround levels in premium movie theaters. These theaters are usually professionally engineered and set up to high standards. If your surround at home is more "immersive" than you're hearing at the movies, think twice.

## Video Tuning

Virtually all modern video displays offer several video parameter presets to assist the user in finding a setup that is most appealing. Even so, most TV pictures you see in the home have excessive color level, brightness and contrast. As these parameters are increased, the picture seems to be enhanced, but is the end result realistic? Real humans don't walk around with a reddish glow. What usually happens is that the user boosts the color first to take advantage of the new TV's color brilliance abilities, then increases the contrast to get a deeper basic picture. This setup then tends to look like it needs more brightness, so that gets a boost. The best approach is probably to start with the "normal" preset, or something that appears like that, and make small adjustments from there after a lot of viewing.

Another setup challenge is movies, particularly dramas comprised mostly of interiors. These shows, including made for

TV movies made to look like movies, are almost invariably dark. Part of it is mood generation, but it's also the fact that these products were designed to be shown in darkened movie theaters. It you juggle your brightness and contrast to make this content show brighter in a lighted room, your other program content, like sports and reality shows, will be over-boosted. The only real solution is to watch this kind of content in a darkened room, and every home media room should have curtains to accomplish this. Besides being technically correct, you'll enjoy the mood crafting that the producers and directors intended.

## Considering the Big Picture

Next, we might consider whether any broad audio frequency response alteration is necessary, as with the receiver or amp bass and treble controls. At this point, any real low end anomalies have probably been reduced or eliminated with changes in speaker placement and/or subwoofer level adjustment.

Today's speakers are so flat to the limits of hearing and dispersion is so good that any treble adjustment is unlikely to be required, unless you just desire a brighter high end regardless of acoustic accuracy. Path loss from the speakers to a listening position 10 feet way is probably less than a dB, depending upon temperature and humidity. As noted previously, movie theater systems are usually equalized to correct for screen loss and path loss, but in a theater that path is likely to be 70 to 100 feet so it's worthwhile to correct the loss.

An interesting conundrum is that home entertainment

systems are expected to accurately reproduce all popular kinds of music and television audio inputs, and these sources are hugely different in character. The most easily replicated sources would be small musical ensembles, or soloists performing with simple accompaniment in a small space. Jazz, most Country and Light Rock fall into this genre. This is much like having the group in your listening room, and the music is usually recorded with that in mind. A good recording played on a high performance system that is adjusted for best fidelity can be an amazing experience.

On the other end of the complexity spectrum are orchestral performances in large venues like Classical. Interestingly, if every instrument were microphoned at close distance, the resulting recording would put the orchestra minus concert hall acoustics in the speakers' near field of your system. This dry recording would be mighty strange indeed. There is no way on earth for most folks to have a listening environment large enough to yield a decent replica of the concert hall sound. With this in mind, the recording engineers include the sound of the venue in the recording, which is a real art decades in the making.

Dense Rock music is a whole other thing. Most people hear rock bands at or near threshold of pain levels in big spaces, either indoor or outdoor. To replicate this experience, you need to play the recording at very high levels in your home (God help the neighbors) and the systems described in this book can easily achieve those levels. Playing this kind of music at lower levels, even using loudness compensation, loses the punch. The high sound pressure levels are part of the performance.

## System Optimization for Source Type

Attempting to "optimize" your home system for any of these sources is probably the wrong approach. In my view and experience, the best "optimization" is to design and adjust for the closest to perfect fidelity your system is capable of. In this way you are presented with the sound that the producers intended, whether it be Rock, Classical, Jazz or Country. Your system will then also be optimum for movie audio reproduction.

Fortunately, graphic equalizers have slowly disappeared from the scene, and that's a good thing. It's probably not that users didn't like them anymore, especially the ones with the neat LED bar graph displays, but that receivers for home entertainment systems with 6 (5.1) or 9 (7.2) channels would need a lot of equalizer channels and displays.

Messing around with response in the midrange frequencies is dangerous as it can totally change the character of the sound. Altering the treble can be easily accomplished with the receiver's treble control. As mentioned previously, with today's speakers, altering the bass response is best left to the subwoofer controls. Except for low frequencies, attempting to correct for "room" response is an exercise in futility. To understand how complex this is, it is useful to look at the nearest professional version of the typical home media room: the recording studio control room. This is hardly a coincidence since the recording engineers and producers need to create product that will sound great in millions of living rooms, family rooms, media rooms and man caves.

## History of Control Room Designs

The basic result of reflections between the surfaces of a room is the generation of eigentones, which are the fundamental resonant frequencies of a room. Parallel surface reflections in a room reinforce the sound having a wavelength of twice that of the dimensions of the room. There are at least three modes of resonance in a room. These frequencies will therefore be reinforced. Early control room designers were aware of room resonances, but did little else in the way of acoustic treatment than apply absorptive material to the walls and ceiling to minimize the resonances. In the 1950s this was not a very scientific pursuit and studio designers were more concerned with the view lines over the consoles and through the window into the studio. The control room was rather empirically determined to "sound good", or not, mostly dependent upon the speaker performance.

The first major real research and implementation of scientifically designed control rooms were those of Tom Hidely in the 1960's. Although he was enamored with the sound of speakers in free space, he realized that control rooms had to deal with practical necessities, such as the room itself and the window to the studio. So, in general, he specified control rooms to provide pretty much near-field response from the speakers to the listening position and dampened the resonances. His control room designs featured:

- Absolute symmetry along a median plane in the room to create a stable stereo image.

- No reflections coming from the back wall.
- No reflections coming from the ceiling.
- Monitor loudspeakers built into and flush mounted with the front wall of the room.
- A short reverberation time of the control room down to and including low frequencies (the 63 Hz-octave band).

Hidely used hanging panels of absorptive material in the rear of the room and on or above the ceiling to get the required low frequency absorption. Although he called them bass traps, they actually absorbed higher frequencies too and the rooms had fairly flat reverberation time over a broad range.

The next major advance in control room design came in the 1980's with the introduction of Time Delay Spectrometry (TDS). TDS, developed by audio science pioneer Richard C. Heyser, allowed separating the direct sound from the speakers from the room reflections so that the near field response at the listening position could be analyzed without room interference. This was critical to measuring an accurate "house curve" to document the control room sound so that any engineer using the room will know its characteristics.

At about the same time, the Live End - Dead End methodology was finding broad use, although it has pretty much disappeared from the scene. This approach deadened the area around the listening position so the engineer was hearing mostly the near-field response of the speakers and made the rear of the room very live so some reverberant energy would be reflected into the listening position.

In the late 1990's **EBU Tech. 3276**, "Listening conditions for the assessment of sound programme material: monophonic and two-channel stereophonic" was introduced with the goal of defining a standard control room environment, however, it has not been very widely applied.

While all of this background is probably more than you ever wanted to know about control room acoustics and practice, I include it for a reason. With all of the research into control room acoustics over the last 50 years, it is still an evolving art despite the efforts of some giants in audio design. The common design goal of these approaches over the years has been to deliver the near field sound of the monitor speakers to the engineer with some reverberant sound from the back of the room to combine with the direct near-field thus simulating a typical listening room at home. Therefore, if you have a system with flat response at the speakers and a reasonably normal room environment, you have what the recording engineers are working towards. Do you really want to attempt to EQ your system?

Well, we've come to the end of our video and audio adventure and I hope you've found it exciting and rewarding. Be sure to review the Q & A that follows. Video and high fidelity audio are wonderful fields that can be enjoyed on almost any scale over a period of many years, and it's not unusual to have young people take it up in their teens and continue to enjoy it into retirement. I hope this book provokes some thought and adds to your knowledge base!

# 8 Q & A

We've covered a lot of territory in our exploration of audio and video and the related cost factors, and some of the concepts presented are necessarily somewhat complex and a number of questions typically arise. The following Q & A is presented in the order of topics covered to make it easy to skip ahead if you need further clarification of some point.

**Q. If excellent quality equipment can be purchased within a carefully constructed budget, what's the point of risking making a very serious mistake by cutting corners to save some money?**

A. If money is truly no object within the constraints of the project budget, then just spending all of the budget is certainly an option. However, if the same audible system performance can be obtained at a savings, and you know what you're doing, there's really no risk of making a serious mistake. The primary purpose of this book is to give you the confidence to make the best purchase decisions. The money saved can probably buy some great music and movies or system accessories.

**Q. Why are "double blind" comparative listening tests required to determine differences in sound quality? Why not just listen to one source and then the other, listening for the differences?**

A. If the differences are large, then simply listening to the sources might disclose the differences. However, audio A/B comparisons of amplifiers, CD players, DVD players and other source devices involve very small differences and biases can be powerful influences. For example, it is natural to expect that a device costing twice as much as the alternative would be the better sounding, but this is not necessarily true. Also if one unit is playing slightly louder than the other it is frequently judged superior to the other. Double blind means that the person selecting the A & B sources and the person listening to them do not know which source is playing. As a practical matter, these days they are usually the same person, so in a way, you are protecting yourself from your own biases.

**Q. Some equipment manufacturers and reviewers poo poo AB testing saying that the superiority of their products only become audible in long-term listening. How would AB testing capture these long-term qualities?**

A. When the audio fidelity of virtually all good quality equipment on the market reached the point that there was little, if any, audible difference between products, enterprising marketing people introduced the notion of ephemeral qualities that could only be heard in long-term listening. The power of suggestion can be very real, and if someone tells you you will hear something, you might, especially since you won't want to be deficient in hearing nuances that others can detect. After all, who can prove you don't hear something? Soon, it got to be cool to talk about things you can only hear in long-term listening and a new audio myth was born.

**Q. The Fletcher Munson and later curves show an almost impossible degree of bass and treble boost at low volume levels to produce subjectively flat response. What's the solution?**

A. There is none. Even if you use full loudness compensation, if available, and turn the bass and treble controls all the way up, at very low volume levels the reproduction will still sound thin. As a practical matter, low levels are usually for background music and thumping bass and piercing treble would be intrusive. A far worse sounding and potentially damaging situation would be to accidentally leave all that loudness compensation on while the volume is increased to higher levels!

**Q. If a system that has 0.1% distortion is considered very clean, wouldn't one that has 0.01% sound even cleaner?**

A. No, because either spec describes a distortion level that is inaudible. A distortion level of 0.1% is 60 dB below, or 1,000th the desired audio signal that is also present., and the masking effect of the audio would make it impossible to detect such tiny distortion components. The generally accepted threshold of Total Harmonic Distortion audibility is 1%, 100th, and that's using pure test tones. With complex music waveforms it would be higher.

**Q. Why didn't the radio broadcast industry adopt the iBiquity digital broadcasting system years ago when it was first introduced and approved by the FCC?**

A. Like many technology introductions, the impediments were technical, economic and political. To enable continued operation of the existing analog system along with the digital signals, only a small portion of the transmitted power is available to the digital. This made the digital signal coverage less than the analog signal, which is anathematic to most broadcasters. And, there were still other issues related to interference, etc. There were other issues related to license fees and equipment costs that were significant expenses, especially for smaller operators. After all, there were no digital receivers, so we had the classic chicken and egg conundrum. Receiver manufacturers didn't want to develop receivers until there were lots of stations to listen to, and broadcasters didn't want the expense of the conversion unless there were lots of listeners with digital receivers. The political angst was the usual multifaceted struggles between iBiquity, the FCC and broadcasters.

**Q. Why is the fidelity of AM radios today worse than 50 years ago?**

A. For AM transmission, the maximum frequency response is a function of receiver bandwidth. The more broadly a receiver tunes, the better the potential high frequency audio response, but the greater the susceptibility of interference, especially from adjacent channels. The narrower the bandwidth the sharper the receiver tunes and the more immune it is to interference. As years have gone by, sources of radio interference have multiplied tremendously. Ever try to listen to a weak AM station in a room with fluorescent lights? As a result, receiver manufacturers have had to design for narrower bandwidth, otherwise consumers bring the sets back as defective. Only a handful have adjustable bandwidth. The other big factor was the introduction of FM with its inherently high fidelity. Over the years music programming went on FM and voice programming went to AM.

**Q. What is pre-emphasis all about?**

A. Pre-emphasis is all around us in the audio world, especially in analog noise reduction systems like Dolby, and it's also part of many digital compression systems. Its effectiveness is based on the fact that most audio has decreasing amplitude vs. frequency. If you have a 1/3 octave bar graph display that is part of an equalizer, for example, you can easily see that the signal amplitude decreases rapidly as the frequency rises. Therefore, the high frequencies can be boosted on recording and then attenuated by a complimentary amount on playback. Flat response results, but any noise on the recording side is reduced because it's usually hiss, which is mostly high frequency. Pre-emphasis is used in AM and FM broadcasting, TV, analog movie sound and many lossy digital compression systems.

**Q. Why can't speakers be made with as low distortion and wide frequency response as amplifiers?**

A. Speakers must convert an electrical input to a mechanical motion equivalent and finally couple the mechanical movement of the speaker diaphragms (cones) to the air. Each step of the electro, mechanical, acoustic conversion process has its inaccuracies. Frankly, it's amazing that the resulting audio from a quality system is as good as it is. The heart of the speaker is the magnet and voice coil, which is in effect a motor. In fact, speaker designers often refer to that assembly as the motor. Over the years, attempts have been made to minimize the mechanical link. One of these is the electrostatic speaker which features a thin flat membrane made to vibrate between two grids charged with high voltage audio, generally produced by a step-up transformer. While these designs generally can achieve very low distortion at mid through high frequencies, bass reproduction is a problem and some electrostatic systems use a conventional driver for the bass. The fully horn-loaded Klipschorn and Klipsch La Scala are 15 dB (over 5X) more sensitive than typical direct radiator speakers, are full range and very clean. They cost several thousand dollars, but owners keep them literally for decades.

**Q. Why are subwoofers non-directional and can they really be placed anywhere in the listening room?**

A. As the frequency of sound waves gets lower the wavelength get longer. By the time the frequencies are below a few hundred Hz, the waves are incredibly long and the human ear gets its directional cues from mid and high frequencies. For example, a 100 Hz tone has a wavelength of almost 1000 feet! However, the closer the listener is to the subwoofer the louder it will sound, and that in itself gives a directional cue. It's a bad idea to place the subwoofer next to the listening position. Anywhere more than a few feet away that produces smooth bass response is fine.

**Q. If applying EQ makes a system sound better, why is that bad?**

A. This gets down to who the artist is. The goal of high fidelity as a pursuit has always been to reproduce the source material as exactly as possible. If adding bass and/or treble makes the reproduction sound "better" to the listener in control there is nothing wrong with that as long as that listener is aware that this is not consistent with high fidelity. In the formative years of the audio art, the extremes of the audio range were lacking. So, when woofers and tweeters revealed those delightful musical nuances, adding even more bass and treble seemed to make a good thing better, and it was a status symbol to have a Hi-Fi set that boomed and sizzled away. It was not unusual to see bass and treble controls at their maximum settings and everyone had their favorite demo records. As equipment improved over the years, especially speakers, along with improved source material, the desire for additional EQ waned as listeners became more sophisticated. Except for attempts to EQ for room abnormalities, you don't hear much about EQ anymore. EQing for room acoustic variations is usually a bad idea because your ear/brain system can hear through the room variations and what you're really doing is warping your system response. Refer to <u>How We Hear</u>. However, if what you end up with sounds better to you than what you started with, give it a try. But listen to a lot of source material to make sure you haven't induced any quirks that affect certain types of music.

**Q. Why does my car system seem to have better fidelity than my big home system, especially the clarity of vocals?**

A. In the car you are in the near fields of the speaker systems. This means that you are hearing essentially what is radiated by the speaker diaphragms without response variances from room reflections and acoustic coupling losses in the bass. This is close to what you hear in headphones, except that most headphones are actually equalized to simulate some typical losses, otherwise they would be too bright to be realistic. Dialog is more intelligible in auto systems and headphones because you're so close to the speakers and there are no smearing effects of multiple room reflections. If you listen to your home system right in front of the speakers you'll hear similar clarity. Car audio is generally not equalized in the treble, except for listener preference. The bass is often boosted to compensate for the masking effect of road noise. Ever notice how bassy your car radio sounds when you're stopped with the engine off and how it gets thinner as your speed increases? Some car radios have adjustable compensation for this, but most folks who have it don't know what it is and don't use it.

---

# GLOSSARY

## 5.1, 6.1, 7.1

Descriptors of multi-channel amplifier specifications to support surround sound and subwoofers. The number before the decimal point is the number of full response audio channels provided. The number after the decimal is the number of low frequency extension or subwoofer channels outputs provided. Example: left, right and center front channels plus left and right surround channels = 5, plus 1 subwoofer output = .1

## AM Stereo

Analog AM Stereo systems were proposed by six manufacturers in the late 1980s. The FCC failed to select a single standard and left the decision to the marketplace. Most broadcasters preferred the Motorola CQAM system. Due to lawsuits by Kahn Communications claiming that Motorola had breached its patents, receiver manufacturers and broadcasters were reluctant to adopt any system and AM stereo faded into obscurity.

## Atmos

Dolby cinema audio processing system that adds a height dimension to surround sound thus allowing movement or placement in three dimensions. Source audio must be Atmos encoded, and theater or home audio must include an Atmos decoder.

## Binaural Audio

Two channel audio recording technique in which the microphones are imbedded in the ears of a dummy head to closely duplicate the human hearing sound path. Playback is usually through headphones to maintain the integrity of the path.

## Blu-ray

DVD format providing full 1080p High Definition Video with Blu-ray encoded DVDs. Most Blu-ray players also provided some degree of "upscaling" to improve the picture quality of standard DVDs.

## Bluetooth

Method of short-range radio transmission in the 2.4-2.485 GHz band originally developed for cellular telephone headsets. Later higher fidelity versions became popular for wireless stereo headsets and current applications include wireless speakers and amplifiers.

## Center Channel

The center channel in movie and home entertainment systems keeps the dialog attached to the screen and improves sound source localization overall.

## Clipping, Clipping Distortion

Flat topping of the audio waveform when an amplifier is called on to exceed its maximum power output.

## Cocktail Party Effect

The ear/brain system's ability to hear through background noise to concentrate on a desired conversation. The same ability allows hearing lecturer speech in large reverberant auditoriums.

## Compression

In analog designs compression is the limiting of dynamic range so that variations in audio levels are eliminated or reduced and the density of the audio is increased. In digital technology complex mathematical algorithms are used to decrease the digital capacity required to store or transmit data. In audio there is lossless and lossy compression. Lossless compression only reduces data where it will not be audible, but the data reduction is modest. Lossy systems, like mp3, further reduce the data but at the expense of some slight change in fidelity.

## Crossover Distortion

In "push-pull" amplifier output stages designed for maximum efficiency there is a finite time in the crossover from one waveform polarity to the other where the two halves of the "push pull" circuit do not exactly match up. This generates crossover distortion and is generally audible only at low levels and was common in early solid state amplifiers.

## dB, Decibel

A logarithmic measure of power or voltage ratios to make comparisons of levels more closely relate to human perception. Examples: In audio, a power ratio of 2:1 equals 3 dB, which would be a 30% change in voltage or sound pressure level. A 4:1 power ratio would equal 6 dB, which would be a 2:1 voltage or sound pressure level change. The dB is also used to compare signal levels in transmission technology. In general, 1 dB is the smallest loudness change that can be detected.

## Dolby

The trade name of a number of signal processing products developed and marketed by Dolby Laboratories. In audio, these are both analog and digital audio noise reduction systems. In video, Dolby Video improves color and contrast for motion pictures and home TV.

## Dynagroove

An LP tracing distortion cancellation technique introduced by RCA in 1963 which faded from the scene by the 1970's. The system pre-distorted the audio recorded on LP records to partially cancel the distortion on playback with common conical styli. Dynagroove also included some dynamic EQ to boost the highs and lows at low signal levels. There was much controversy about Dynagroove at the time because its benefit was aimed at making inexpensive players sound better at the expense of premium audiophile equipment with full response and employing elliptical styli.

## Eigentone

The fundamental resonant frequency of a room. Parallel surface reflections in a room reinforce the waves of sound having a wavelength of twice that of the dimensions of the room. There are at least three modes of resonance in a room. These frequencies will therefore be reinforced.

## Elliptical Stylus

A premium stylus design developed to reduce tracing distortion on playback of LP records. Tracing, or pinch effect distortion, was the result of the inability of normal conical shaped styli to trace recorded waveforms in the narrowest part of the record groove.

## Equalization, EQ

Alteration of frequency response in a device or system, usually to correct for frequency response errors elsewhere in the signal path.

## Fletcher-Munson Curves

Average audio response of human hearing at various audio levels from the threshold of hearing to the threshold of pain introduced by researchers Harvey Fletcher and Wilden Munson in 1933. ISO226 in 2003 showed general conformity with the Fletcher-Munson curves from 1933.

## Frequency Response

The relative levels of equipment amplification or human hearing loudness perception versus frequency. The smallest change the human ear can perceive is 1dB, or about 10% change in sound pressure level.

## Harmonic

In music, audio and radio transmission technology a multiple of a base frequency. Example: 800 Hz is the second harmonic of 400 Hz.

## HDR

In high definition television technology HDR is High Dynamic Range. This is not to be confused with the HDR photographic process. For TV, HDR increases the range of contrast from black to maximum brightness and the range of colors that can be reproduced. HDR improves the naturalness of TV displays.

## Hz, Hertz

In an alternating current or sound pressure waveform, the number of cycles per second frequency, named after Heinrich Rudolf Hertz in 1960.

## Halo Effect

In audio, hearing an effect that doesn't exist because one believes that it does because of assumed causative factors. For example, hearing a warmer sound from vacuum tube amplifiers because the tubes are glowing hot.

## Hi Resolution

In audio and video, fineness of detail that is greater than normal or traditional. In audio this is generally reproduction accuracy greater than CD quality. In video it is generally picture resolution of 720 lines or greater with 1080 lines progressive being the current standard.

## IMD

Intermodulation distortion, which is the result of a device or system nonlinearity causing two or more signals to interact with each other, thus generating non-harmonically related distortion. In audio, this type of distortion is very irritating.

## iBiquity

A company formed by the merger of USA Digital Radio and Lucent Digital Radio and now owned by DTS. The iBiquity system is an in-band-on-channel (IBOC) digital broadcasting scheme that allows digital broadcasting along with standard analog AM and FM, or all digital broadcasting. The system was originally developed in response to US broadcaster demands that no digital system be implemented that would result in more radio stations being licensed and therefore more competition.

## Integration

In audio, the change in waveform shape due to the effects of filtering. The human ear/brain system also performs integration when presented with sound with a frequency exceeding its high frequency capability. The audible result is that there is no difference in the sound of a complex waveform comprised of harmonics above the audible range and a single frequency sound at the limit of audibility. For example, a 7.5 kHz square wave sounds the same as a 7.5 kHz sinewave.

## Klipschorn

A large horn-loaded speaker system designed for corner placement so that the floor and walls present a continuation of the horn's acoustic gain. First introduced by speaker design pioneer Paul W. Klipsch in 1948, the electrical to acoustic conversion efficiency is 40 times greater than normal speakers.

## LFE, Subwoofer

Low frequency extender, later called subwoofer, is a specially designed speaker system, either passive or amplified, usually with internal equalization, to enable reproduction of very low frequencies. Since low frequencies are non-direction due to their long wavelength, a subwoofer may be placed anywhere in a room, and one subwoofer may be used to extend the response of any number of channels.

## Loudness Compensation

Boosting low and high frequencies in accordance with the equal loudness compensation curves to compensate for human hearing losses at low loudness levels. Some amplifiers and receivers have compensation controls to provide either fixed or automatic variable compensation.

## MTS

Multichannel Television Sound. The analog TV stereo audio technology adopted in 1984.[11] Created by the Broadcast Television Systems Committee, the stereo TV system was originally named BTSC. It enabled modulating the TV audio subcarrier with an audio difference channel (L-R), which when added to the L+R mono in a receiver matrix reproduced the stereo audio from the TV or Cable station.

## mp3

Mpeg Layer III digital encoding that reduces digital audio file size by an average of 90%. To achieve this degree of digital reduction the system employs lossy compression that results in a loss of some audio fidelity. Extensive listening tests during and after the development of mp3 demonstrated that most listeners could not hear the losses. Mp3 is used for most digital music downloads and streaming.

## Node

In acoustics, room nodes are the result of reflections from the surfaces of the listening room. These occur in such a way as to produce peaks and dips in the reproduction of audio from speakers as the reflections cancel and reinforce in numerous locations.

## OLED

Organic light emitting diode. Enables pixels which can achieve full black level for maximum contrast from display without backlighting. A TFT backplane is employed to drive the pixels directly. OLED displays also exhibit wider viewing angles and faster response time.

## Pink Noise Generator

A white noise generator filtered to produce a spectrum similar to music. Pink noise generators are frequently used as test signal source in amplifiers and receivers for user testing of amplifier channels and correct operation of the speakers. The tonality of pink noise will noticeably change if a woofer, tweeter or midrange drive in a speaker system is out.

## RF Sensitivity

In audio, the sensitivity of radio devices such as tuners, Bluetooth devices, remote controls, etc. More broadly, the sensitivity of any radio receiving device.

## SPL

Sound pressure level, usually stated in dBs. The threshold of human hearing is 0dB and the threshold of pain is 120 dB.

## Stereo Image

When presented with a true stereo input, two adequately separated speakers will produce a continuous soundstage, including a phantom center channel. A convincing stereo image requires good amplifier or tuner stereo separation, good speaker radiation characteristics, a good room environment and a well centered listening position.

## Subwoofer, LFE

A specially designed speaker system, either passive or amplified, usually with internal equalization, to enable reproduction of very low frequencies. Since low frequencies are non-directional due to their long wavelength, a subwoofer may be placed anywhere in a room and one subwoofer may be used to extend the response of any number of channels.

## THD

Total harmonic distortion. The sum of all harmonically related spurious components generated by a device or system nonlinearity. Humans can detect harmonic distortion starting at about 1%. Excellent audio equipment will have less than .1% THD.

## THX

An audio quality assurance standard developed by George Lucas's company in 2002 for application in motion pictures. Since then it has been used for just about everything with audio, from home audio equipment, computers, software and video games. The unmistakable audio signature preceding every THX certified event is the glissando starting from a very low note, which came to be called *Low Note*.

## Tracing Error

Tracing, or pinch effect distortion, was the result of the inability of normal conical shaped styli to trace recorded waveforms in the narrowest part of the record groove. Tracing error distortion increases with frequency and is greatest on the inside grooves of a record.

## Ultra HD

4K television, which has over 4 times the resolution of 1080p HD has been dubbed Ultra HD. Ultra HD also offers improved color characteristic and contrast dynamic range.

## Upscaling

Video upscaling improves the picture quality of sources with less than HD quality when they are played on HD and Ultra HD devices. The most common upscaling is from Blu-ray DVD players with upscaling to HD and Ultra HD TVs. While upscaling cannot improve the resolution to the full native resolution of the HD or Ultra HD device, the improvement can be very noticeable.

# BIBLIOGRAPHY

1. Warren, Richard M. *Auditory Illusion and their relation to Mechanisms Normally Enhancing accuracy of Perception.* Journal of the Audio Engineering Society, September 1983.

2. Deutsch, Diana. *Auditory Illusions, Handedness, and the Spatial Environment*. Journal of the Audio Engineering Society, September 1983.

3. Davis, H. *Hearing,* John Wiley and Sons, New York 1938.

4. Hobbs, Marvin. *E. H. Scott... The Dean of DX, A History of Classic Radios.* North Frontier Press 1985.

5. Ciapura, Dennis R. *The More Things Change...A product from the past reigns as the ultimate in AM receiver Technology.* Broadcast Engineering, February 1990.

6. Fletcher, Harvey. *Symposium on Wire Transmission of Symphonic Music and Its Reproduction in Auditory Perspective.* Electrical Engineering January 1934.

7. Various, Wikipedia, *Stereophonic Sound.* Last updated March 25, 2017.

8. Vilchur, Edward. *Problems of Bass Reproduction in Loudspeakers.* Journal of the Paul W. Audio Engineering Society, July, 1957.

9. Klipsch, Paul W. *Corner Speaker Placement.* Journal of the Audio Engineering Society, July, 1959.

10. Ciapura, Dennis R., *Audio Fidelity, The Grand Illusion.* Recording Engineer/Producer, November 1989.

11. Ciapura, Dennis R., *The Real World of Stereo TV.* Broadcast Engineering, September 1986.

12. Heyser, Richard C. *Concepts in the Frequency and Time Domain Response of Loudspeakers,* Proceedings of IREE, March, 1976.

# THE TRUTH ABOUT STEREO AND VIDEO SPECS

# INDEX

5.1 29, 59

AM Stereo 17, 59, 60

Blu-ray 40, 43, 49, 60, 69

Binaural 18, 60

Bluetooth 31, 32 60, 67

Compression 12, 13, 14, 18, 21, 32, 56, 61, 66

Crossover Distortion 22, 23, 61

Clipping 23, 61

Dolby 18, 19, 56, 62

Dynagroove 20, 62

Elliptical Stylus 19, 62

Equalization, EQ 38, 39, 57, 58, 62, 63, 65, 68

Fletcher Munson Curves 8, 53

HDR 42, 63

Harmonic Distortion, THD 9, 10, 11, 14, 16, 54, 68,

Halo Effect 24, 64

Intermodulation Distortion, IMD, TIMD 11, 12, 64,

iBiquity 13, 14, 17 54, 55, 64

Integration 10, 65

Klipschorn 27, 57 65

LFE 28, 65, 68

Loudness Compensation 26, 54 33, 54, 65, 66

MTS 31, 66

MP3 13, 14, 21, 31, 32, 47, 61, 66

OLED 42, 43, 66, 67

Pinch Effect 20, 62, 68

Pink Noise Generator 47, 67

RF Sensitivity 16, 67

Subwoofer 6, 9, 28, 29, 34, 35, 37, 39, 42, 44, 45, 46, 47, 48, 49, 50
      57, 59, 65, 68

SPL 27, 30, 67

Stereo Image 42, 67

Tracing Error 20, 68

Ultra HD 44, 68, 69

Upscaling 60, 69

## ABOUT THE AUTHOR

Dennis Ciapura has spent most of his career in broadcasting and professional audio and has written over 80 articles for the Broadcasting and Recording Engineering trade journals. He has built everything from crystal sets to recording studios and radio stations. His area of special interest includes psycho-acoustics and he has been an outspoken critic of pseudo-scientific theories of audio reproduction. Dennis lives in Montana and when not critiquing the latest theory of super audio is likely to be found fishing in a nearby lake.

Made in the USA
Middletown, DE
02 October 2022